ସଂଖ୍ୟାର କାହାଣୀ

THE NUMBER STORY

SMALL BOOK ONE

ENGLISH - ODIA

*Numbers Teach Children
Their Number Names*

written and illustrated by

MISS ANNA

Early Reader Edition of *The Number Story 1*
Bronze Medal Winner, 2016 Wishing Shelf Book Award

Library of Congress Control Number: 2018902040

Names: Miss Anna, author.
Title: Number story : numbers teach children their number names / Miss Anna.
Description: Portland, OR: Lumpy Publishing, 2018.
Identifiers: ISBN 978-1-945977-91-6 | LCCN 2018902040
Summary: The pictures and rhymes present stories which introduce numbers 0-10.
Subjects: LCSH Numeration—English--Odia--Pictorial works--Juvenile literature. | BISAC JUVENILE NONFICTION /
Languages: English--Odia
Classification: LCC QA141.3 .M57 2018 | DDC 513—dc23

Publisher: Lumpy Publishing
Website: www.missannabooks.com
Email: missanna@missannabooks.com
Facebook: Miss Anna Lumpy

Paperback: ISBN 978-1-945977-91-6
Printed in the U.S.A.   1 3 5 7 9 10 8 6 4 2

କ'ଣ ତମ୍ଭେମାନେ ସଂଖ୍ୟାମାନଙ୍କର ନାଆଁ ଶିଖିବାକୁ ଚାହିଁବ?

It is very easy and a lot of fun!

ଏହା ବହୁତ ସହଜ ଆଉ ବହୁତ ମଜାର!

Say-along our little jingle

ଆମ ସହିତ ଆମ ଛୋଟ କାହାଣୀଟି ଗାଅ ।

starting from Number One!

ଆମ୍ଭେ ସଂଖ୍ୟା ଏକ ଠାରୁ ଆରମ୍ଭ କରିବା।

1

ONE looks like my one finger.

୧ ଏକ

ଏକ ମୋର ଗୋଟିଏ ଅଙ୍ଗୁଠି ଭଳିଆ ଅଟେ।

1
ONE!
এক!

2

TWO   trails a tail.

୨. ଦୁଇ

ଦୁଇ ଗୋଟିଏ ବଙ୍କା ଲାଞ୍ଜ।

A TAIL! ଲାଞ୍ଜଟି!

3

THREE has bumps.

ଓ  ତିନି

ତିନି ର ବମ୍ପ ଅଛି।

BUMPY!

ଅସମାନ!

4

FOUR carries a sail.

୪ ☆ ଚାର

ଚାରି ର ଗୋଟିଏ ନୌକାପାଲ ଅଛି ।

A SAIL!
ଯାଲଟି!

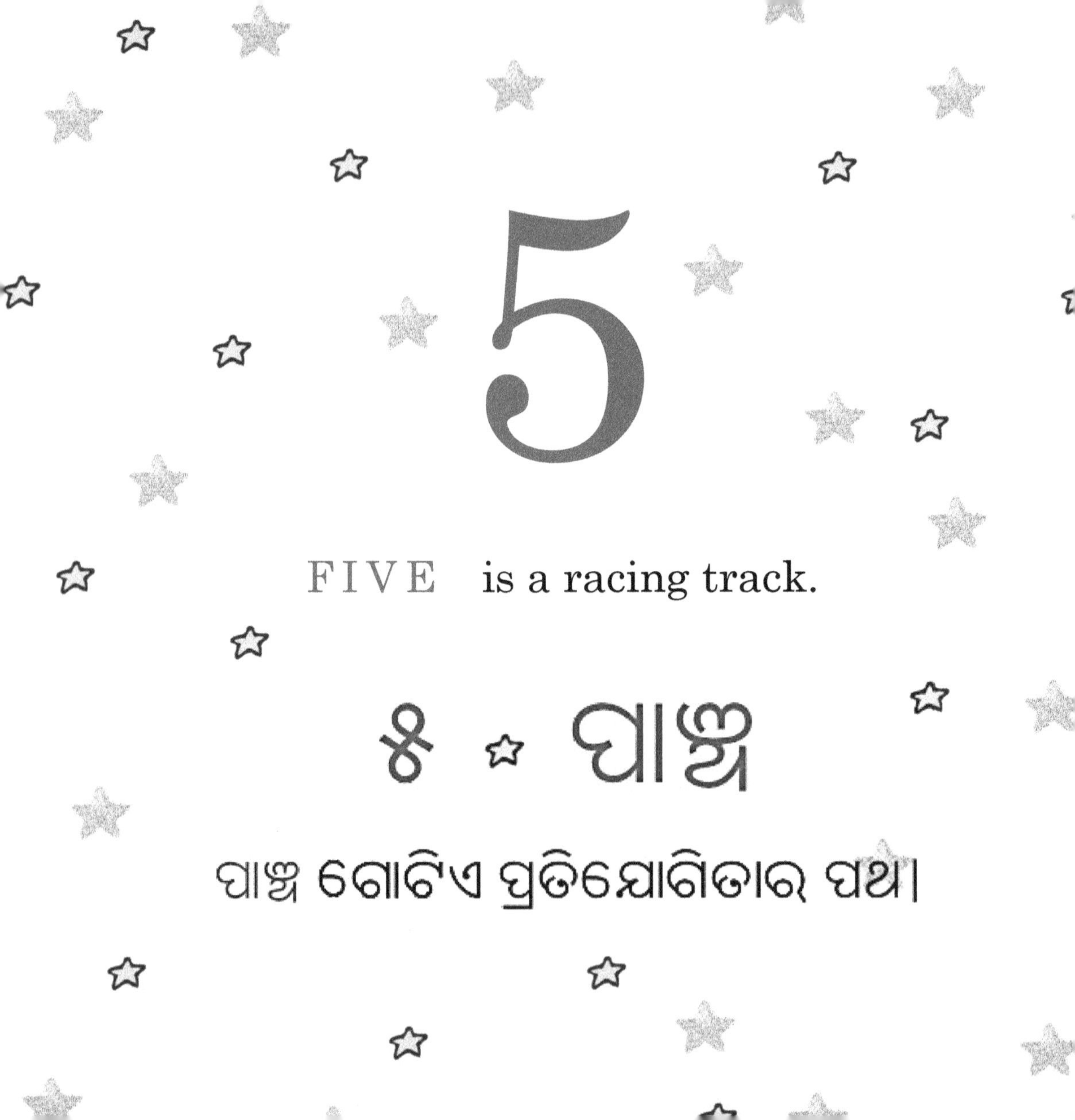

5

FIVE is a racing track.

୫ ⭐ ପାଞ୍ଚ

ପାଞ୍ଚ ଗୋଟିଏ ପ୍ରତିଯୋଗିତାର ପଥ।

VROOM VROOM!

6

S I X   curves like a snail.

୬ ☆ ଛଅ

ଛଅ ଗୋଟିଏ ଗେଣ୍ଡୁ ଭଳିଆ ବକ୍ର।

A SNAIL! ଗେଣ୍ଡୁଟି!

7

SEVEN has a sharp angle.

୭ ⋆ ସାତ

ସାତ ର ଗୋଟିଏ ଧାରୁଆ କୋଣ ଅଛି।

OUCH!
ଆଉଚ୍!
BE CAREFUL! IT'S SHARP!
ସାବଧାନ! ଏହା ଧାରୁଆ!

8

୮ ଆଠ

ଆଠ ଗୋଟିଏ ରୋଲାରକୋଷ୍ଟାର ରେଲ ଅଟେ।

ଓଡ଼ିଶା!
YIPPEE!

NINE   is a bubble on a stick.

୯ ⋆ ନଅ

ନଅ ଗୋଟିଏ ବାଡ଼ି ଉପରେ ଫୋଟକାଟି।

A BUBBLE! ଫୋଟକାଟି!

10

TEN is an eye of a whale.

୧୦ ଦଶ

ଦଶ ଗୋଟିଏ ତିମିର ଗୋଟିଏ ଆଖି।

WINK!
ଆଖିମାର!

HELLO! ହେଲୋ!

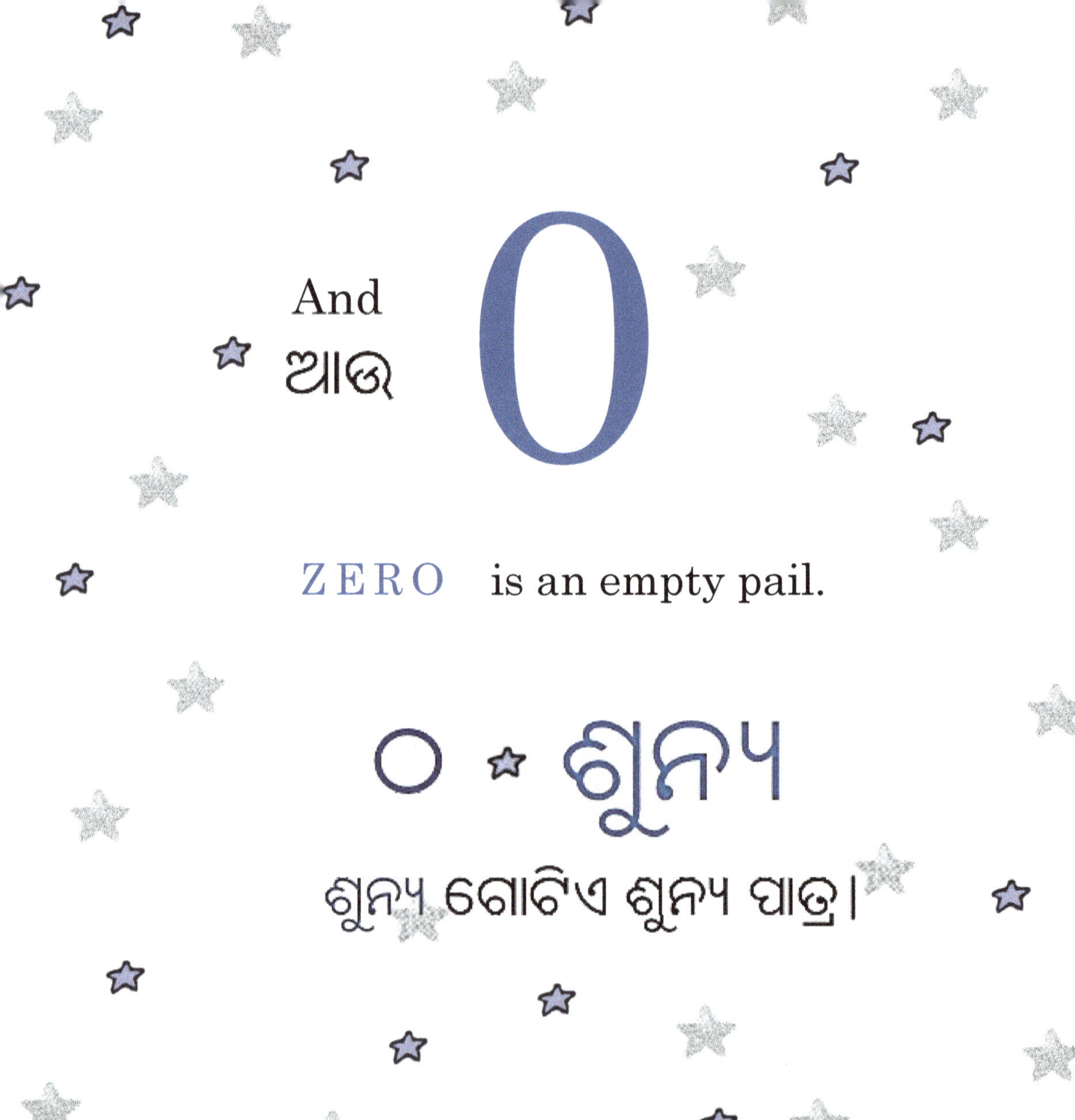

And
ଆଉ

0

ZERO   is an empty pail.

○ ⋆ ଶୂନ୍ୟ

ଶୂନ୍ୟ ଗୋଟିଏ ଶୂନ୍ୟ ପାତ୍ର।

IT'S
EMPTY!
ଏହା ଶୂନ୍ୟ!

Thank you for playing with us today.

We had a lot of fun too!

ଆଜି ଆମ ସହିତ ଖେଳିବା ଯୋଗୁ ତମ୍ମାନଙ୍କୁ ଧନ୍ୟବାଦ ।
ଆମେ ମଧ ବହୁତ ମଜା କଲୁ !

We are your Number friends,
Zero to Ten,
Who will be here for you~

ଆମ୍ଭେ ତମର ସଂଖ୍ୟା ମିତ୍ର
ଶୂନ୍ୟ ରୁ ଦଶ।
ଆମ୍ଭେ ଏଠାରେ ସବୁବେଳେ ତମ୍ଭମାନଙ୍କ ପାଇଁ ରହିବୁ।

Bye-bye now!
See you again soon!

ଏବେ ପାଇଁ ବିଦାୟ!
ପୁଣି ବହୁତ ଶିଘ୍ର ଦେଖାହେବ!

The Numbers are *SINGING* too!

To sing-a-long, look for Miss Anna Number Story
at your favorite music store like iTUNES.

MP3

Numbers 0-10
IDENTIFYING
& COUNTING

Numbers 11-20
& Ordinals

first, second, third...

Numbers 0-100
& Place Values

ones, tens, hundreds...

About Clocks
& Telling Time

hours, minutes, seconds

Number Story 1 & 2

isbn: 978-0-996216-48-7

Number Story 3 & 4

isbn: 978-1-945977-01-5

Number Story 5 & 6

isbn: 978-1-945977-06-0

Number Story 7 & 8

isbn: 978-1-949320-40-4

For more Miss Anna books to love,
visit us at

www.missannabooks.com

Numbers are working hard all over the world!
Come Travel the World with Us!